Editor
Lorin E. Klistoff, M.A.

Editorial Manager
Karen J. Goldfluss, M.S. Ed.

Editor-in-Chief
Sharon Coan, M.S. Ed.

Cover Artist
Janet Chadwick

Art Coordinator
Kevin Barnes

Art Director
CJae Froshay

Imaging
Rosa C. See

Product Manager
Phil Garcia

Publishers
Rachelle Cracchiolo, M.S. Ed.
Mary Dupuy Smith, M.S. Ed.

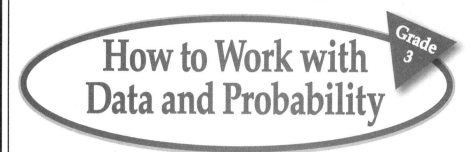

How to Work with Data and Probability

Grade 3

Author

Mary Rosenberg

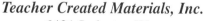

Teacher Created Materials, Inc.
6421 Industry Way
Westminster, CA 92683
www.teachercreated.com

ISBN-0-7439-3739-2

©2003 Teacher Created Materials, Inc.
Made in U.S.A.

The classroom teacher may reproduce copies of materials in this book for classroom use only. The reproduction of any part for an entire school or school system is strictly prohibited. No part of this publication may be transmitted, stored, or recorded in any form without written permission from the publisher.

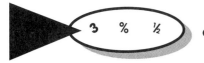

Table of Contents

How to Use This Book 3
A Note to Teachers and Parents 3
About This Book 3
NCTM Standards 4

Unit 1: How to Categorize Events 5
What Are the Chances? 6
What's for Lunch? 7
Making a Graph to Show the Data 8

Unit 2: How to Work with Sample Sizes 9
One Red Jellybean 10
More Than One Red Jellybean 11
Adding Even More Jellybeans 12

Unit 3: How to Organize Data 13
Reading a Graph 14
Making a Graph 15
Making a Bar Graph 16

Unit 4: How to Collect Data 17
Interviewing 18
Observing 19
A Taste Test 20

Unit 5: How to Make Different Kinds of Graphs 21
A Line Graph 22
Pie Charts 23
A Bar Graph 24

Unit 6: How to Analyze the Data 25
The Mean 26
The Median 27
The Mode 28

Unit 7: How to Work with Probability and Dice 29
Rolling a Die 30
Rolling a Pair of Dice (Part A) 31
Rolling a Pair of Dice (Part B) 32

Unit 8: How to Work with Probability and Spinners 33
It's About the Spinner 34
Wheels of Chance 35
Is it Likely? Is it Unlikely? 36

Unit 9: How to Work with Probability and Playing Cards 37
It's All in the Suits 38
It's All in the Deck of Cards 39
More Than One Suit 40

Unit 10: Brain Teasers
Too Many Choices 41
Even More Choices 42

Unit 11: Technology
Tables and Graphs 43

Unit 12: Challenge
It's All in the Dots 44
Double-Line Graph 45
Go on a Scavenger Hunt 46

Answer Key 47

 **Use This Book**

A Note to Teachers and Parents

Welcome to the "How to" math series! You have chosen one of several books designed to give your children the information and practice they need to acquire important concepts in specific areas of math. The goal of the "How to" math books is to give children an extra boost as they work toward mastery of the math skills established by the National Council of Teachers of Mathematics (NCTM) and outlines in grade-level scope-and-sequence guidelines.

This book is intended to be used by teachers and parents for a variety of purposes and needs. Each of the individual units contains one or more "How to" pages and two or more practice pages. The "How to" section of each unit precedes the practice pages and provides needed information such as a concept or math rule review, important terms and formulas to remember, or step-by-step guidelines necessary for using the practice pages. While most "How to" pages are written for direct use by the children, in some lower-grade level books, these pages are presented as instructional pages or direct lessons to be used by a teacher or parent prior to introducing the practice pages.

About This Book

How to Work with Data and Probability: Grade 3 introduces the concepts of probability and data analysis to the first-time learner, then extends these concepts to the relationship between data analysis and probability and how they apply to the real world.

The activities in this book will help your children learn important new skills or reinforce skills already learned in the following areas:

- Determining the likelihood of outcomes, both in words and as quantities expressed as percents, decimals, and fractions
- Creating and analyzing different kinds of tables and charts to determine possible outcomes
- Analyzing data to predict outcomes
- Identifying independent and dependent variables relating to probability
- Calculating the mean, median, mode, and range
- Collecting, organizing, and analyzing data

Each subsequent concept builds upon the learning of the previous skills. Teachers and parents are encouraged to complete the book in the order as it is presented here, unless students are in need of remedial instruction in specific areas.

Regardless of their ability to add, subtract, multiply, and divide, students may complete the practice pages following the concepts presented on "How to" pages with ease. Students may use a calculator when working with higher numbers unless their computation skills are in need of remediation.

Once having completed this book, students will feel like professional statisticians ready to apply their learning to all areas of their lives.

 **Standards**

How to Work with Data and Probability: Grade 3 matches a number of NCTM standards in the following areas:

Statistics

The activities in this book allow students to explore statistics in real-world situations. This includes the collection, organization, and description of data, and the construction and interpretation of charts, tables, and graphs. Students also must make practical and sensible decisions based on the analysis of data and provide arguments for their decisions.

Probability

The activities in this book provide model situations in which students must determine probabilities and express results in several ways. Students must also make predictions based on probable experiment and theory, and justify decisions based on results.

Computation and Estimation

The activities in this book offer a wide array of open-ended, real-world problems for which students must use their mathematical expertise to provide solutions. As part of the problem-solving process, students meaningfully apply their problem-solving skills to verify and interpret results and justify decisions based on these results.

Communication

Students are given numerous opportunities to apply physical diagrams, charts, graphs, tables, tally sheets, pictures, and materials to concrete mathematical ideas. Students can relate their everyday common language to the expression of mathematical ideas and symbols on a level appropriate to their age. Students will also understand that mathematics involves discussion, reading, writing, and listening as integral functions of mathematical instruction.

Reasoning

Students learn to apply logic to their math problems and to justify their answers. There is an emphasis on the use of models, manipulatives, and charts. Students learn to apply rational thinking effectively and correctly to problems rooted in real-life circumstances.

Connections

Students are encouraged to recognize and relate various mathematical concepts, processes, and patterns to each other. They are likewise encouraged to use mathematics across the curriculum and in their daily lives.

Other Standards

The activities in this book provide a range of additional skill application such as mathematical reasoning and proportions. This book also provides connections to other subject areas and to the world outside the classroom, including the identification and interpretation of patterns and functions through the use of charts, tables, and graphs.

 •••••••••••••••• **Categorize Events**

The first step in working with probability is categorizing events based upon prior experiences or known information. An event can be categorized as *impossible*, *unlikely*, *equal chance*, *likely*, or *certain*.

Facts to Know

- **Event:** An occurrence or outcome
- **Percent:** An event occurring can be expressed as a percentage of one.

    Example: If a coin is tossed, what is the chance of it landing on heads?

    Answer: 1 out of 2 or 50% or half the time

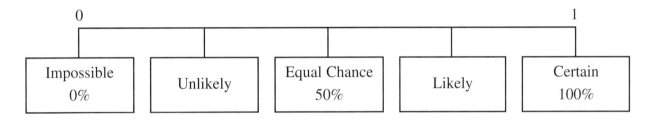

- **Impossible:** Zero chance the occurrence or outcome will happen

    Example: The moon being made of green cheese (*Impossible*—This is impossible because through samples collected by astronauts, scientists know that the moon is not made of green cheese.)

- **Unlikely:** Less than a 50% chance the occurrence or outcome will happen

    Example: It has only rained in June three times over the last ten years. What is the chance that it will rain in June this year? (*Unlikely*—Rain is not impossible, because it has rained in June in the past. There is a small chance that rain might occur but it probably won't. Therefore, rain in June is unlikely.)

- **Equal Chance:** A 50-50 chance the occurrence or outcome will happen or will not happen

    Example: Brenda has made goals at five of the last ten games. What are the chances that Brenda will make a goal at tonight's game? (*Equal Chance*—Brenda has made goals in half the games. Her chances are equal that she might make a goal, or she might not make a goal, at tonight's game.)

- **Likely:** More than a 50% chance the occurrence or outcome will happen

    Example: Brad's spelling scores are 90%, 100%, 95%, 96%, 98%, and 75%. What are the chances that Brad will earn a score of at least 90%? (*Likely*—Brad will likely earn a score of at least 90% based on earning at least 90% on five out of the six spelling tests.)

- **Certain:** Positively sure the chance the occurrence or outcome will happen

    Example: If the Beaverton Beavers baseball team has won 100 out of the last 100 games, what is the team's chance of winning the next game? (*Certain*—The team has won every game. It is certain the team will win the next game.)

1 Practice — What Are the Chances?

Directions: What are the chances of each event occurring for you today or happening to you today? Place a check mark under the correct category for each event.

	Impossible	Unlikely	Equal Chance	Likely	Certain
1. The sun rising					
2. Riding a bike					
3. Climbing Mt. Everest					
4. Losing a tooth					
5. Driving a car					
6. Finding an undiscovered Van Gogh painting					
7. Meeting the president					
8. Riding in a limousine					
9. Watching the cow jump over the moon					
10. Attending a school event					
11. Seeing a clown					
12. Reading a book					
13. Playing a game					
14. Seeing a rainbow					
15. Building a snowman					
16. Playing in the rain					
17. Going to school					
18. Writing a letter					
19. Calling a friend					
20. Flying to the moon					

1 Practice — What's for Lunch?

Directions: Use the lunch menu to answer each question.

Monday	Tuesday	Wednesday	Thursday	Friday
Pizza	Chicken	Hot Dog	Baked Potato	Hamburger
Pizza	Nachos	Salad	Soup	Hamburger
Pizza	Sandwich	Taco	Baked Potato	Hamburger
Pizza	Corn Dog	Spaghetti	Soup	Hamburger

What are the chances of having the following items on next month's lunch menu?

1. Soup on Tuesday?
   Likely   Unlikely
2. Hamburgers on Friday?
   Likely   Unlikely
2. Pizza on Thursday?
   Likely   Unlikely
4. Corn dogs on Wednesday?
   Likely   Unlikely
5. Spaghetti on Monday?
   Likely   Unlikely
6. Tacos on Wednesday?
   Likely   Unlikely
7. Nachos on Friday?
   Likely   Unlikely
8. Sandwiches on Tuesday?
   Likely   Unlikely
9. Baked potatoes on Thursday?
   Likely   Unlikely
10. Soup on Thursday?
    Likely   Unlikely
11. Salad on Friday?
    Likely   Unlikely
12. Chicken on Tuesday?
    Likely   Unlikely
13. Hot dogs on Thursday?
    Likely   Unlikely
14. Hamburgers on Tuesday?
    Likely   Unlikely

1 Practice — Making a Graph to Show the Data

Directions: What is the chance of . . .

Key		
U = Unlikely	E = Equal Chance	L = Likely

_____ 1. Pulling the Queen of Hearts out of a deck of cards?

_____ 2. Rolling six 1's in a row?

_____ 3. A quarter landing on "heads" five out of ten tosses?

_____ 4. A penny landing on "tails" ten times in a row?

_____ 5. Catching a ball ten times in a row without dropping it?

_____ 6. Receiving a $2 bill as change at three different stores?

_____ 7. Getting five green lights in a row?

_____ 8. Finding five out of ten people with the same birthday?

_____ 9. Winning every game in Bingo?

_____ 10. Winning at least one game out of ten in Bingo?

_____ 11. Being able to kick a rolling soccer ball?

_____ 12. Getting a green light at two out of four intersections?

_____ 13. Picking a green lollipop out of a pair of lollipops?

_____ 14. Receiving change from buying a toy?

_____ 15. Drinking a glass of water sometime during the day?

_____ 16. Getting the mail?

_____ 17. Finding a penny on the ground?

Directions: A graph can be made to show how each statement was answered. Color one box for each statement's answer.

	1	2	3	4	5	6	7	8	9	10
Unlikely										
Equal Chance										
Likely										

Directions: Write a sentence about the graph.

_____

2 How to Work with Sample Sizes

Children need to develop the understanding on how the sample population can skew the results either negatively or positively.

Facts to Know

- **Sample Size:** The number of items in the testing group in relation to the total number or population also known as the *data set*.

  ✔ The smaller the percentage of the sample population in relation to the whole population equals results that are less reliable and difficult to replicate.

  ✔ The larger the percentage of the sample population in relation to the whole population equals results that are reliable and easy to replicate.

Look at the following examples of how results can be skewed by the people or items involved in the sample populations.

   a. Over one-half of the people overwhelmingly pick salt-free soda as their favorite drink. (Sample population consisted of two people.)

   b. 50% of the people say water is their favorite drink. (Sample population size was 50,000.)

   c. 5% of the students will volunteer to work in their communities. (Results were based upon interviewing 10,000 students aged 12–15.)

   d. 20% of the students will not volunteer to work in their communities. (Results were based upon interviewing one 12-year-old student.)

   e. Pollsters predict only 30% of the eligible population will vote in the local election. (Results were based upon telephone interviews of 1,000 people who may or may not be eligible to vote.)

   f. Pollsters predict only 30% of the eligible population will vote in the local election. (Results based upon telephone interviews of 1,000 people who are eligible to vote.)

- **Event:** An occurrence or outcome—a turn or a selection

- **Experiment:** A test of some sort

- **Accurate:** The reliability of the results or findings

- **Reliability:** If the results were accurate, the test can be repeated by other people and achieve similar results.

2 Practice • • • • • • • • • • • • • • • • • • One Red Jellybean

Directions: In a bag, place one red jellybean or one index card labeled with the word *red*. Take the jellybean (or card) out of the bag ten times. Record the results from each event.

Red										
Orange										
Yellow										
Green										
Blue										
Purple										
Black										
Pink										
White										
	1	2	3	4	5	6	7	8	9	10

Directions: Answer the following questions using the results above.

1. Which color of jellybean was picked the most often? _____

2. Why was this jellybean picked the most often? _____

3. If another person repeated this experiment, would the results be the same? Why or why not? _____

4. Would the results be the same if other colors of jellybeans were added to the bag? Why or why not? _____

5. Would this experiment show an accurate representation of the colors of jellybeans in a bag? Why or why not? _____

6. What could be done to make this experiment more accurate? _____

 **More Than One Red Jellybean**

Directions: In a bag, place one jellybean of each color (or one index card labeled with each color). Take one jellybean (or card) from the bag, record the color on the chart, and then return the jellybean to the bag. Repeat this procedure ten times and then answer the questions below the graph.

	1	2	3	4	5	6	7	8	9	10
Red										
Orange										
Yellow										
Green										
Blue										
Purple										
Black										
Pink										
White										

1. Which color of jellybean was picked the most often? _____
2. Which color of jellybean was picked the least often? _____
3. Were any color of jellybeans picked the same number of times? If yes, what colors?
   _____
4. If this experiment were repeated, would the results be similar? Why or why not?
   _____

Directions: Repeat the experiment and compare the results with the graph at the top.

	1	2	3	4	5	6	7	8	9	10
Red										
Orange										
Yellow										
Green										
Blue										
Purple										
Black										
Pink										
White										

© Teacher Created Materials, Inc.

 **Practice** •••••••• **Adding Even More Jellybeans**

Directions: Place in the bag five jellybeans of each of the listed colors or five labeled index cards for each jellybean color. Answer the question below.

1. Which color of jellybean do you think will be picked the most often? Why?

_____

Directions: Take one jellybean (or card) from the bag, record the color on the chart, return the jellybean to the bag. Repeat this procedure ten times and then answer the questions below.

Red										
Orange										
Yellow										
Green										
Blue										
Purple										
Black										
Pink										
White										
	1	2	3	4	5	6	7	8	9	10

2. Which color of jellybean was picked the most often? _____

3. Which color of jellybean was picked the least often? _____

4. Were there any colors of jellybeans picked the same number of times? If yes, what colors? _____

5. If this experiment were repeated by another person, would the results be similar? Why or why not? _____

_____

6. Which sampling method would be the most accurate—selecting from one jellybean, from nine jellybeans, or from 45 jellybeans? Why? _____

_____

3 How to .... Organize Data

How the data is organized can effect how useful it is to other readers. Graphs that are clearly labeled and presented in a logical manner make it easier for others to read and understand the information. Well-designed graphs also make it easy to quickly find specific information about one of the items. They also can identify weaknesses, strengths, number of participants/items used in the study, to track results or trends or some other needed information.

Facts to Know

(Use the graph below as a point of reference.)

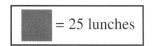

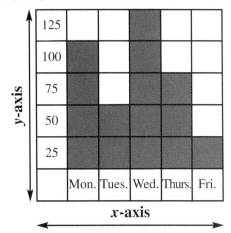

- **Title:** The title lets the reader know what information is being presented in the graph. For this graph, the title is "Hot Lunches Ordered for This Week."

- **Scale:** The scale tells what is being measured and how it is being measured. There are two scales. One scale is along the y-axis (the vertical axis). On this graph, the scale along the y-axis tells how many lunches were ordered in units of 25. The other scale is along the x-axis (the horizontal axis). On this graph, the scale along the x-axis tells how many hot lunches were ordered each day.

- **x-axis:** The x-axis is horizontal and is located at the bottom of the graph. The items on the x-axis are read going from left to right.

- **y-axis:** The y-axis is vertical and is located on the left side of the graph. The items on the y-axis are read going from the bottom to the top.

- **Range:** The range tells the difference between the highest and lowest numbers. For this graph, the high is 125 points and the low is 25 points. The range is $125 - 25 = 100$ points.

- **Legend:** The legend lets the reader know some important points about the graph. For this graph, the legend says that each shaded box is equal to 25 lunches.

3 Practice • • • • • • • • • • • • • • • • • • Reading a Graph

Directions: Identify the parts of the graph by answering the questions below.

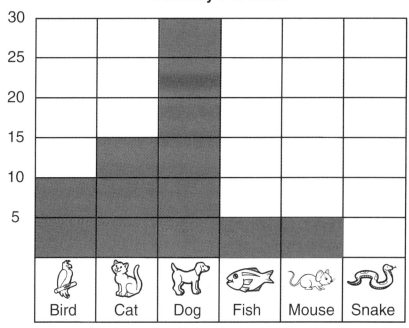

1. The *scale* tells what is being measured and how it is being measured. What are the scales for this graph? _____

2. The *x-axis* is horizontal and is located at the bottom of the graph. The *x*-axis is read going from left to right. What information is shown on the *x*-axis? _____

3. The *y-axis* is vertical and is located on the left side of the graph. The *y*-axis is read going from the bottom to the top. What information is shown on the *y*-axis? _____

4. The *legend* provides important information about the value of each shaded space. What information does the legend on this page tell about the graph? _____

5. The *range* tells the difference between the highest number and the lowest number. What is the range for this graph? _____

Directions: Circle the questions that can be answered by reading the graph.

6. How many pets were sold on Monday?
7. Are there more dogs or more cats?
8. Does the Friendly Pet Store have any snakes?
9. How many people bought a bird yesterday?
10. Which pet is Marcie's favorite?
11. How many animals are there in all?
12. Are there more cats and mice or dogs and birds?

3 Practice — Making a Graph

Directions: Stacy asked 15 classmates about their hobbies. Use tally marks to organize the data.

Allen	baseball	Frankie	swimming	Kevin	swimming
Bea	skiing	George	skiing	Laura	baseball
Cass	swimming	Helene	baseball	Mike	skiing
Derikka	baseball	Ivan	skiing	Naomi	swimming
Eddie	baseball	Justina	skiing	Oliver	baseball

1. Make one tally mark for each classmate's hobby. Write the total number for each hobby.

	Tally Marks	Total Number
Baseball		
Skiing		
Swimming		

2. The tally marks can be used to create a bar graph. Make a bar graph to show each classmate's hobby. Color one box to show each classmate's favorite hobby.

3. Write a statement about the graph.

_____

_____

4. Write a question that can be answered by the information shown on the graph.

_____

_____

	Baseball	Skiing	Swimming
10			
9			
8			
7			
6			
5			
4			
3			
2			
1			

© Teacher Created Materials, Inc.

3 Practice · · · · · · · · · · · · · · · · · · Making a Bar Graph

Directions: Use the information to make a bar graph. The students had one minute to make as many stars as possible. Below are the results.

Paul	50 stars	Tom	50 stars	Wanda	100 stars
Queenie	90 stars	Ursula	60 stars	Xavier	90 stars
Roberto	80 stars	Vince	90 stars	Yvonne	50 stars
Susie	80 stars				

Step 1: Write the names of each student along the *x*-axis. Writing the names in alphabetical order makes it easier for the reader to find the information.

Step 2: Write the number of stars along the *y*-axis. To find the range, the highs and lows, put the numbers in numerical order first.

_____, _____, _____, _____, _____, _____, _____, _____, _____, _____

Then write the numbers in order starting at the bottom of the *y*-axis with the lowest number and moving up with the remaining numbers. If there are two or more scores with the same number, just write the number one time on the graph. Since the information for this graph uses tens, make sure you do not leave any of the numbers out—even if nobody had made that specific number of stars!

Step 3: Shade the boxes to show each student's score.

Step 4: Write three questions that can be answered using the information shown on the graph.

_____

_____

_____

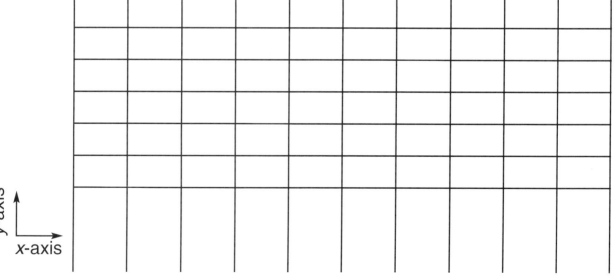

Stars Made in One Minute

Collect Data

Data, or information about the world around us, can be collected in a variety of ways. Some of the ways are very familiar to most people.

Facts to Know

- **Tally marks:** A way of counting items using lines (1's) and "bundles" of lines (5's)

  | = 1 || = 2 ||| = 3 |||| = 4   ̶|̶|̶|̶|̶ = 5   ̶|̶|̶|̶|̶ ̶|̶|̶|̶|̶ = 10

  ̶|̶|̶|̶|̶ ̶|̶|̶|̶|̶ | = 11   ̶|̶|̶|̶|̶ ̶|̶|̶|̶|̶ || = 12   ̶|̶|̶|̶|̶ ̶|̶|̶|̶|̶ ||| = 13

- **Interview:** Asking people how they feel about a certain product or topic
- **Taste Test:** Having people taste two or more unlabeled items
- **Comparing Results:** Telling how two or more outcomes are alike and/or different
  Examples:
  ✔ Most of the girls picked the green bikes over the purple bikes.
  ✔ Most boys between ages 5–8 picked soccer as their favorite sport.
  ✔ The third graders liked the green apples better than the red apples.
  ✔ All of the students like to eat hot lunch, but the younger students like pizza the best. The older students like the corn dogs the best.
- **Product:** An item such as a toy, game, article of clothing, food item, etc.
- **Test Subject:** Person participating in the experiment
- **Observation:** Watching and recording events as they happen—It is like being a "silent witness."
- **Findings:** The result or the outcome of the experiment
- **Trends:** Pattern to the findings

The Bike Shop's Sales of Mountain Bikes

Examples using the information from the graph on the right:

✔ Girls between ages 7–10 bought more mountain bikes than any other age group.
✔ Boys of all age groups bought mountain bikes in the same numbers.
✔ Younger girls and older girls bought very few mountain bikes.
✔ Overall, kids between ages 7–10 bought the most mountain bikes.

© Teacher Created Materials, Inc.

4 Practice — Interviewing

Directions: Select a product for 10 people to test. The product can be a food item, a game, a new toy, a piece of furniture, a pencil, or even a video. Color a box to show how each person voted. Write down why each person did or did not like the product.

Item: _____

Yes										
No										

Yes, I liked the product because:	No, I didn't like the product because:
_____	_____
_____	_____
_____	_____
_____	_____
_____	_____

Summarize the findings. Tell the main reasons why people liked and/or disliked the product.

_____
_____
_____

Note any trends by answering the following questions:

1. Who seemed to like the product the best?

    boys   girls

2. Which age group liked the product the most?

    2–6 years   7–12 years   13+ years

3. What could be done to make the product more appealing to more people?

    _____
    _____

4 Practice • Observing

Directions: Use tally marks to keep track of the vehicles that pass by for ten minutes (or five minutes if there is a lot of traffic).

	Day: _____ Time: _____	Day: _____ Time: _____
Motorcycles		
2-Door/4-Door Cars		
Vans, SUV's, Station Wagons		
Trucks		
School Buses		

1. Of which kind of vehicle were there the most? _____

2. Of which kind of vehicle were the fewest? _____

3. Would you see more specific kinds of vehicles at certain times of day? If yes, what kind of vehicle and why? _____

4. If you kept track of vehicles again on a different day but at the same time, do you think the results would be the same? Why or why not? _____

5. Keep track of vehicles again but on a different day or at a different time. Compare the results. Were there any similarities or differences?

6. If you were to watch the traffic again, of what other information could you keep track?

4 Practice — A Taste Test

Directions: Select a product for people to test. Some product ideas include the following:

- two different types of apples
- two different kinds of bread
- two different types of milk
- diet soda and regular soda
- two different brands of the same kind of cookie
- two different kinds of ice cream

When conducting the experiment, make sure the test subjects do not know the brands (or kinds) of food items they are testing. Put small samples of each item on a paper plate or in small cups and label one paper plate "Sample A" and the other paper plate "Sample B." At the end of the experiment, share the results with the test subjects. Note any comments the test subjects made regarding the products.

Product Items:

Sample A = _____

Sample B = _____

Sample A										
Sample B										

Test Subjects' Comments: _____

_____

_____

1. Which product did the test subjects like the most? Why? _____

_____

2. Which product did the test subjects like the least? Why? _____

_____

3. Were the results what you had expected? Why or why not? _____

_____

4. Were the test subjects surprised by the choices they made? Why or why not? _____

_____

5. Were any trends noticed in how each test subject voted? (Examples: Did more males (or females) like one product more than the other? Did younger people prefer one product over the other product?) _____

_____

5 How to ...... Make Different Kinds of Graphs

Information about a topic can be shown on a graph. There are many different kinds of graphs such as line graphs, bar graphs, and pie charts. Which graph to use depends upon the information being represented.

Facts to Know

- **Pie Chart:** A pie chart shows information for one specific category. Each "slice" of the pie chart represents one piece of information. (See the chart on the right.) By looking at the chart, the reader can see how each "slice" relates to the other slices and to the total "pie."

    The larger the slice, the greater the number in that particular category.

    The smaller the slice, the fewer the number in that particular category.

    Examples of when to use a pie chart include the following:
    1. To show the different categories of how a specific amount of money is spent
    2. To show the favorite hobbies for a specific number of people
    3. To show the top five most popular pets bought in one year

- **Line Graph:** A line graph is used to show change over time. Below is a line graph showing the number of pennies found on the ground over a five-day period.

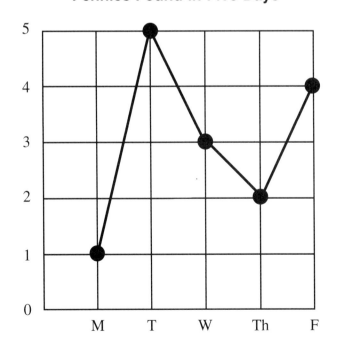

- **Bar Graph:** A bar graph is used to show specific amounts.

    Examples of when to use a bar graph include the following:
    1. To show the total yield of crops
    2. To show the top five favorite restaurants as voted on by a specific population
    3. To show the record sales for different recording artists
    4. To show the record sales for different albums
    5. To show the number of students in each grade (See the bar graph below.)

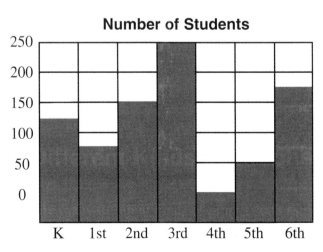

© Teacher Created Materials, Inc. 21 #3739 How to ... Data and Probability: Grade 3

5 Practice · A Line Graph

Directions: A *line graph* is used to show change over a period of time. A dot is made on the graph to show specific information at a specific time. Once all of the events have been plotted (marked on the graph) the dots are connected. Use the information below to make a line graph.

Bea's Pet Walking Business

Customers

January:	10 pets	May:	4 pets	September:	6 pets
February:	6 pets	June:	9 pets	October:	3 pets
March:	4 pets	July:	1 pet	November:	3 pets
April:	7 pets	August:	3 pets	December:	6 pets

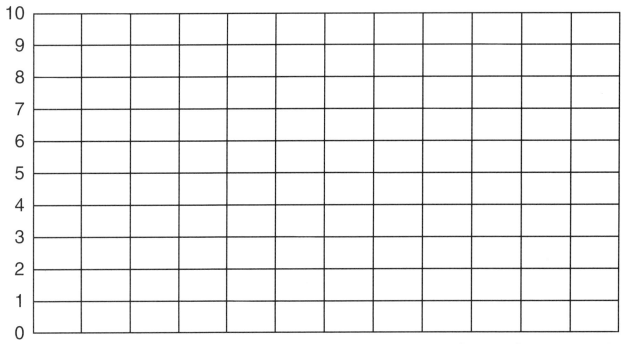

1. Write a sentence about the information shown on the graph.

_____

_____

2. Write a question that can be answered by using the information shown on the graph.

_____

_____

5 Practice • Pie Charts

A *pie chart* is used to show the different pieces of information relating to one specific category. Each piece is a "slice" of the whole. (Think of a pizza cut into slices of different sizes.)

Directions: Daniel receives a weekly allowance of $8.00. Complete the pie chart to show how Daniel spends his allowance.

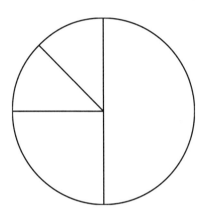

Daniel's Allowance
$4.00 for a movie ticket
$1.00 for pet food
$2.00 for snacks
$1.00 saved

Directions: Find the percentage for each "slice" of money. (Example: Amount spent on pet food. $1.00 ÷ $8.00 = .125 x 100 = 12.5%)

1. Amount spent on snacks: _____

2. Amount saved: _____

3. Amount spent on a movie ticket: _____

4. In addition to labeling the pie chart, what could be added to make the information stand out to the reader? _____

Directions: If Daniel's allowance increased to $10.00 a week, how would the information shown on the pie chart change? Create a new pie chart using the same percentage of his allowance for each category.

Daniel's Allowance
5. ____% or $_____ for a movie ticket
6. ____% or $_____ for pet food
7. ____% or $_____ for snacks
8. ____% or $_____ saved

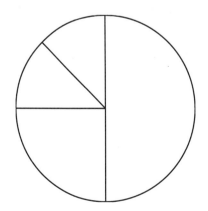

© Teacher Created Materials, Inc. 23 #3739 How to ... Data and Probability: Grade 3

5 Practice — A Bar Graph

Directions: A *bar graph* is used to show information at a given time. Make a bar graph showing each vehicle's miles per gallon (mpg). (If a vehicle's mpg falls in between two values, shade the box to show the approximate mpg.)

Miles Per Gallon for Different Kinds of Vehicles

Vehicle	mpg	Vehicle	mpg
Compact Car	24 mpg	Scooter	35 mpg
Limousine	19 mpg	Sedan	20 mpg
Motorcycle	33 mpg	SUV	14 mpg
Pickup	17 mpg	Tractor	6 mpg
Recreational Vehicle (RV)	10 mpg	Van	15 mpg

[Bar graph grid with vehicles (Compact Car, Limo, Motorcycle, Pickup, RV, Scooter, Sedan, SUV, Tractor, Van) on the y-axis and mpg values (0, 5, 10, 15, 20, 25, 30, 35, 40, 45 mpg) on the x-axis.]

1. Write a sentence about the information shown on the graph.

_____

_____

2. Write a question that can be answered by looking at the graph.

_____

_____

6 How to • • • • • • • • • • • • • • • • • • • Analyze the Data

Statistical analysis is a way of looking at the data through different measurements called the *mean*, *median*, and *mode*.

Facts to Know

- **Mean:** The mean is the average. To find the mean, add together all of the numbers within the group and divide by the total number of items in the group.

    Example: 10, 3, 4, 2, 8, 17

    10 + 3 + 4 + 2 + 8 + 17 = 44

    44 ÷ 6 (the number of items in the set) = 7.33

- **Median:** The median is the middle number. To find the median, order the numbers within the group from smallest to largest. If there are an odd number of numbers within the group, the median is the middle number. If there are an even number of numbers within the group, the median is found by adding together the two middle numbers and dividing by two.

    Example 1: Odd number of numbers

    5, 15, 39, 3, 53

    Order the numbers: 3, 5, (15,) 39, 53   The median is 15.

    Example 2: Even number of numbers

    29, 46, 98, 10, 88, 92

    Order the numbers: 10, 29, (46, 88,) 92, 98

    Add together the two middle numbers: 46 + 88 = 134

    Divide by two: 67.   The median is 67.

- **Mode:** The mode is the number (or numbers) that occur the most often within a given set of numbers.

    Example: 13, 36, 79, 35, 25, 82, 36, 64, 72, 36, 79

    The mode is 36. It occurs three times within this set of numbers.

- **Range:** The range is the difference between the smallest number and the largest number.

    Example: 10, 5, 47, 6, 54, 14, 68

    The smallest number is 5 and the largest number is 68.

    68 − 5 = 63   The range is 63.

- **Ordering:** To arrange items into a particular sequence such as from smallest to greatest, greatest to smallest, shortest to tallest, tallest to smallest.

- **Odd:** Number that cannot be evenly divided by two. There will always be one item left. Odd numbers have a 1, 3, 5, 7, or 9 in the ones place.

- **Even:** Number that can be evenly divided by two. There will not be any items left. Even numbers have a 0, 2, 4, 6, or 8 in the ones place.

- **Scales:** Tell what is being measured and how it is being measured.

    The scales are located on the *x*-axis and the *y*-axis.

    (See the example to the right.)

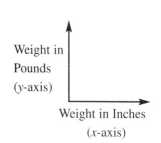

Weight in Pounds (*y*-axis)

Weight in Inches (*x*-axis)

6 Practice ································ The Mean

Directions: The *mean* is the average. To find the mean, add all of the numbers on the list together and divide by the total number of items on the list.

Neighborhood Paper Routes			
Abby	87 papers	Fred	98 papers
Ben	21 papers	Gabby	105 papers
Cassie	14 papers	Hal	15 papers
Deandre	66 papers	Inez	81 papers
Erica	56 papers	James	38 papers

1. What is the total number of newspapers delivered? _____
2. How many paper carriers are there? _____
3. What is the range? _____
4. What is the mean (average) of newspapers delivered? _____
5. Which paper carriers deliver more newspapers than the mean? _____
6. Which paper carriers deliver fewer newspapers than the mean? _____
7. Which kind of graph can best represent this information—a line graph, a pie chart, or a bar graph? Why? _____

Directions: Make the graph to represent this information.

8. What are the scales used on this graph? _____
9. Write a sentence telling about the graph. _____
10. Write two questions that can be answered by using the information shown on the graph.
_____
_____

6 Practice — The Median

Directions: The *median* is the middle number. To find the median, place all of the numbers in order from smallest to greatest.
- If there are an odd number of numbers, the middle number is the median.
- If there are an even number of numbers, the median is found by adding the two middle numbers together and dividing by two.

Georgia's Pennies

Year	Pennies	Year	Pennies	Year	Pennies
1994	956 pennies	1998	427 pennies	2001	725 pennies
1995	4,110 pennies	1999	776 pennies	2002	327 pennies
1996	5,510 pennies	2000	1,033 pennies	2003	921 pennies
1997	692 pennies				

1. What was the fewest number of pennies saved during the last ten years? _____

2. What was the greatest number of pennies saved during the last ten years? _____

3. Write the numbers in order from smallest to greatest.
____; ____; ____; ____; ____; ____; ____; ____; ____; ____

4. What is the range? _____

5. What is the median? _____

6. Which kind of graph can best represent this information—a line graph, a pie chart, or a bar graph? Why? _____

Directions: Make the graph to represent this information.

7. What are the scales used on this graph? _____

8. Write a sentence telling about the graph. _____

9. Write two questions that can be answered by using the information shown on the graph.

6 Practice — The Mode

Directions: The *mode* is the number of items that occur the most often within a given set of numbers. Answer the questions using the information below.

Each box contains two dozen (24) donuts. Each box has the following kind and number of donuts:

Glazed	6 donuts
Old Fashioned	4 donuts
Chocolate	10 donuts
Plain	4 donuts

1. What is the mode? _____
2. What is the range? _____
3. Which kind of graph can best represent this information—a line graph, a pie chart, or a bar graph? Why? _____

Directions: Make the graph to represent this information.

4. What are the scales used on this graph? _____
5. Write a sentence telling about the graph. _____
6. Write two questions that can be answered by using the information shown on the graph.

7 How to ...... Work with Probability and Dice

Probability is making a decision about a topic based on mathematical data and known information.

Facts to Know

- **Probability:** The chance that there will be a specific outcome for an event

- **Prediction:** A guess based on known information

- **Percent:** This is a mathematical way of expressing the outcome for a specific event. To find the percent when working with a fraction, divide the numerator (the top number) by the denominator (the bottom number) and multiply by 100.

    Example: What is the chance of rolling a six with one six-sided die?

    The chances are 1 out of 6 or 1/6.

    1/6 or 1 ÷ 6 = .166 x 100 = 16.6%

- **Possible Outcomes:** All the possible choices for an event

    Example 1: What are the possible outcomes for flipping a coin?

    There are two possible outcomes, heads and tails.

    Example 2: What are the possible outcomes for Jeannette when running in a race?

    There are three possible outcomes. Jeannette could finish first, last, somewhere in the middle.

    Example 3: What are the possible outcomes for making $10.00 using paper bills?

    There are eleven possible outcomes. The possible outcomes are the following:

    1 - $10.00 bill
    2 - $5.00 bills
    1 - $5.00 bill, 2 - $2.00 bills, and 1 - $1.00 bill
    1 - $5.00 bill, 1 - $2.00 bill, and 3 - $1.00 bills
    1 - $5.00 bill and 5 - $1.00 bills
    5 - $2.00 bills

    4 - $2.00 bills and 2 - $1.00 bills
    3 - $2.00 bills and 4 - $1.00 bills
    2 - $2.00 bills and 6 - $1.00 bills
    1 - $2.00 bill and 8 - $1.00 bills
    10 - $1.00 bills

- **Dependent Events:** Dependent events depend upon the outcome of the previous events.

    Example: There are ten socks in a bag. Five of the socks are black. Three of the socks are red. One sock is blue. One sock is white. What is the chance of pulling a white sock out of the bag?

    Event 1: 1/10 (There is a one out of ten chance of pulling a white sock out of the bag.)

    Event 2: 1/9 (The chance increases because one sock has been removed from the bag.)

    Event 3: 1/8 (The chance of pulling a white sock out of the bag increases as more socks are removed from the bag.)

- **Independent Events:** Independent events are not changed because of earlier events.

    Example: The chances of rolling a "4" on a die remains the same no matter how many times the die is rolled. It does not matter if a "4" has been rolled before. The chance will always be 1/6 or 1 out of 6 numbers.

 **Rolling a Die**

Directions: *Probability*, or chance, is the likelihood of having a specific outcome. To find the probability for a specific action, the possible outcomes need to be known.

1. What are the possible outcomes for rolling a die? A die would land on the following numbers: ____, ____, ____, ____, ____, ____

2. What are the odds for rolling a specific number? _____

3. What is the percentage for rolling a specific number? _____

4. Is each event (roll of the die) dependent or independent of the outcome for the previous rolls? _____

5. Make a prediction. On what number do you think the die will land the most often? _____

Directions: Roll the die 20 times. Color a box to show the outcome of each roll.

6																				
5																				
4																				
3																				
2																				
1																				

6. What was the outcome? _____ Which number was rolled the most often? _____

7. Was your prediction correct? _____

8. If this experiment was repeated, do you think the outcome will change? Why or why not? _____

Directions: Repeat this experiment.

6																				
5																				
4																				
3																				
2																				
1																				

9. Was the outcome the same as before? _____

10. Was your prediction (#8) correct? _____

7 Practice · · · · · · · · · · Rolling a Pair of Dice (Part A)

Directions: *Probability*, or chance, is the likelihood of having a specific outcome. To find the probability for a specific action, the following possible outcomes need to be known.

1. What are the possible outcomes for rolling two dice? The dice could land on the following number combinations:

Possible Outcomes											
Die #1	Die #2	Die #1	Die #2	Die #1	Die #2	Die #1	Die #2	Die #1	Die #2	Die #1	Die #2
___	___	___	___	___	___	___	___	___	___	___	___
___	___	___	___	___	___	___	___	___	___	___	___
___	___	___	___	___	___	___	___	___	___	___	___
___	___	___	___	___	___	___	___	___	___	___	___
___	___	___	___	___	___	___	___	___	___	___	___
___	___	___	___	___	___	___	___	___	___	___	___

2. What are the odds for rolling a specific number combination? _____

3. What is the percentage for rolling a specific number combination? _____

4. Is each event (roll of the dice) dependent or independent of the outcomes of the previous rolls? _____

5. Make a prediction. On what combination of numbers do you think the dice will land the most often? _____

Directions: Roll the dice 20 times. On the following page, color a box to show the outcome of each roll. (*Hint:* Make a small mark on one of the dies to designate one as "Die #1." The other die will be "Die #2.")

6. What was the outcome? Which pair of numbers was rolled the most often?

_____

7. Was your prediction correct? _____

8. If the dice were rolled one hundred times, would one combination of numbers be rolled significantly more times than any other possible combination of numbers? Why or why not? _____

7 Practice — Rolling a Pair of Dice (Part B)

Directions: Color the appropriate box to show each roll of the dice.

Roll															
1-1															
1-2															
1-3															
1-4															
1-5															
1-6															
2-1															
2-2															
2-3															
2-4															
2-5															
2-6															
3-1															
3-2															
3-3															
3-4															
3-5															
3-6															
4-1															
4-2															
4-3															
4-4															
4-5															
4-6															
5-1															
5-2															
5-3															
5-4															
5-5															
5-6															
6-1															
6-2															
6-3															
6-4															
6-5															
6-6															

8 How to Work with Probability and Spinners

Spinners are used in many board games. The results of each spin are completely independent of all previous results.

Facts to Know

- **Spinner:** A spinner is used in many board games and games of chance. A spinner is usually round with multiple spaces. Each space on the spinner might contain a number, color, shapes, etc. Each spin of the spinner is completely independent of all other spins. To make an arrow, hold a paper clip in place with a pencil and give the paper clip a "thump" with a finger.

- **Probability:** Chance of a certain outcome

- **Prediction:** A guess based upon known information

- **Fraction:** A part of the whole
    Example: A pizza is sliced into eight slices. Fernando eats three slices.
    Fernando ate 3/8 of the pizza.

- **Percent:** It shows a part of the whole. To find the percentage of a fraction, divide the numerator by the denominator and multiply by 100.
    Example: Fernando ate 3/8 of the pizza.
    $3 \div 8 = .375 \times 100 = 37.5\%$

- **Equal Chance or Fair Chance:** There is a 50% probability of an event having a certain outcome. The arrows have an equal chance of landing on red or any other color represented on the spinner.

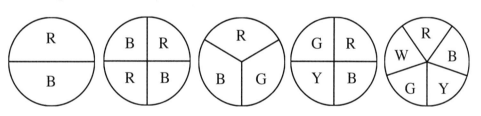

R = red
B = blue
G = green
Y = yellow
W = white

- **Unlikely or Impossible Chance:** There is a less than 50% probability of an event having a certain outcome. The arrows have a less than equal chance of landing on red.

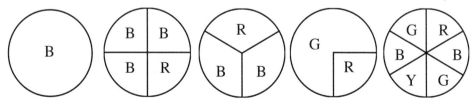

- **Certain or Likely Chance:** There is a greater than 50% probability of an event having a certain outcome. The arrows have a greater than equal chance of landing on red.

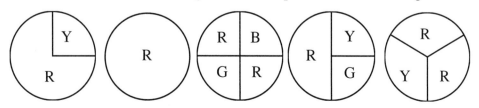

8 Practice • • • • • • • • • • • • • • • • • It's About the Spinner

Directions: Look at each spinner. On which color is the arrow most likely to land? Why? Write your prediction and spin each spinner ten times. Color a box to show the color on which the arrow landed.

1. I think this spinner will land on _____.

(spinner: half red, half blue)

									Red
									Blue

This spinner landed the most often on _____. My prediction was _____.

2. I think this spinner will land on _____.

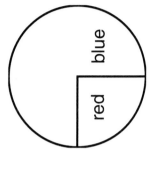

									Red
									Blue

This spinner landed the most often on _____. My prediction was _____.

3. I think this spinner will land on _____.

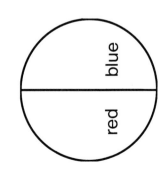

									Red
									Blue

This spinner landed the most often on _____. My prediction was _____.

8 Practice • • • • • • • • • • • • • • • • • • Wheels of Chance

Directions: What are the chances of the arrow landing on red? Write the probability two ways—as a fraction and as a percent. (R = red, Y = yellow, G = green, and B = blue)

1.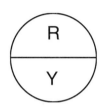

Fraction: _____
Percent: _____

2.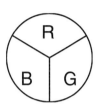

Fraction: _____
Percent: _____

3.

Fraction: _____
Percent: _____

4.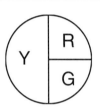

Fraction: _____
Percent: _____

5.

Fraction: _____
Percent: _____

6.

Fraction: _____
Percent: _____

7.

Fraction: _____
Percent: _____

8.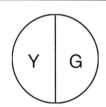

Fraction: _____
Percent: _____

9.

Fraction: _____
Percent: _____

10. Which spinners would have a 50% chance (or more) of landing on red?
_____

11. Which spinners would have less than a 50% of landing on red? _____

12. Which spinners would be fair? _____ Why? _____

13. Which spinners would not be fair? _____ Why? _____
_____

© Teacher Created Materials, Inc.

8 Practice — Is it Likely? It is Unlikely?

Directions: Each spinner was spun ten times. The data under each spinner shows the number of times the arrow landed on each color. How likely is the outcome? Circle the answer. (R = red, Y = yellow, G = green, and B = blue)

1. 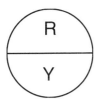 Red: 8 Blue: 0 Green: 0 Yellow: 2 Likely Unlikely	**2.** 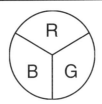 Red: 3 Blue: 3 Green: 4 Yellow: 0 Likely Unlikely	**3.**  Red: 5 Blue: 3 Green: 2 Yellow: 0 Likely Unlikely
4. Red: 4 Blue: 0 Green: 4 Yellow: 2 Likely Unlikely	**5.**  Red: 0 Blue: 8 Green: 1 Yellow: 1 Likely Unlikely	**6.**  Red: 2 Blue: 2 Green: 3 Yellow: 3 Likely Unlikely
7.  Red: 2 Blue: 3 Green: 3 Yellow: 2 Likely Unlikely	**8.** 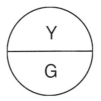 Red: 0 Blue: 0 Green: 5 Yellow: 5 Likely Unlikely	**9.**  Red: 4 Blue: 2 Green: 2 Yellow: 2 Likely Unlikely
10.  Red: 3 Blue: 3 Green: 1 Yellow: 3 Likely Unlikely	**11.** 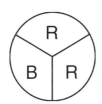 Red: 5 Blue: 5 Green: 0 Yellow: 0 Likely Unlikely	**12.** 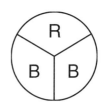 Red: 3 Blue: 7 Green: 0 Yellow: 0 Likely Unlikely

Work with Probability and Playing Cards

Playing cards can provide a concrete experience in working with probability. The activities require the student to use one suit of cards, two suits of cards, and the whole deck of cards.

Facts to Know

- **Consecutively Numbered:** This is when cards are rarranged in numerical (or sequential) order. For playing cards, the sequence is Ace (or 1), 2, 3, 4, 5, 6, 7, 8, 9, 10, Jack, Queen, King.

- **Outcome:** The possible end result of an event
    Example: Marvin has a math test on Friday. What are the possible outcomes?
    1. Marvin will do well on the test.   3. Marvin will not pass test.
    2. Marvin will do okay on the test.

- **Impossible:** A specific outcome will not happen
    Example: Out of a full deck of cards, circle the outcome that is not possible.
    1. A card with a value of ten will be selected.
    2. (A card with a value of thirteen will be selected.)
    3. A odd numbered card will be selected.

- **Unlikely:** A less than 50% chance of achieving a specific outcome
    Example: Which card will have an unlikely chance of being selected? Why?
    1. A face card
    2. A numbered card
    3. (An ace) (Because there are only four aces in a deck of cards. There are twelve face cards and thirty-six cards numbered two to ten.)

- **Equal Chance:** A 50-50 chance of achieving a specific outcome
    Example: Which card will have an equal chance of being selected?
    1. A card with a red heart   3. (A red card)
    2. An ace of hearts

- **Likely:** A more than 50% chance of achieving a specific outcome
    Example: Which card will have a more than equal chance of being selected?
    1. A face card   3. (A numbered card)
    2. An ace of clubs

- **Certain:** A 100% chance of achieving a specific outcome.
    Example: Which card will be selected for sure?
    1. An odd-numbered card   3. (A card of any suit)
    2. A face card

- **Fractional Form:** The odds of a specific outcome can be expressed in a fractional from.
    Example: In one suit of cards, what is the likelihood of selecting a face card? 3/13
        (In one suit of cards, there are thirteen cards from ace to king. Three of the cards are face cards.)

- **Prediction:** An educated guess based on known information

© Teacher Created Materials, Inc.   37   #3739 How to ... Data and Probability: Grade 3

9 Practice — It's All in the Suits

Directions: Take one suit of cards out of the deck of playing cards. Figure out the probability for selecting specific cards.
- The Ace is worth one point.
- The face cards (Jack, Queen, King) are worth ten points.

1. What are the possible outcomes when selecting a card from the suit of cards?
   _____
   _____
   _____

Directions: Write *impossible*, *unlikely*, *equal chance*, *likely*, or *certain* on the line.

2. Picking an ace: _____
3. Picking a face card: _____
4. Picking a numbered card: _____
5. Picking a two: _____
6. Picking an odd-numbered card: _____
7. Picking an even-numbered card: _____
8. Picking two cards that are numbered consecutively: _____

Directions: Which event would be more likely to happen? Circle the event.

9. Picking a lucky number | Picking a card with a value of ten
10. Picking a card with a value greater than eight | Picking a card with a value less than eight
11. Picking a face card | Picking an odd-numbered card
12. Picking an even number | Picking the ace
13. Picking an even-numbered card | Picking an odd-numbered card
14. Picking a card with a value of five | Picking a card with a greater value than five

Directions: Write the odds in fractional form and as a percentage for selecting the following cards.

	Fractional Form	Percentage
15. A face card	_____	_____
16. A number card	_____	_____
17. An odd number	_____	_____
18. An even number	_____	_____
19. A card greater than five	_____	_____
20. A card less than five	_____	_____
21. A card between three and ten	_____	_____
22. A card worth two points	_____	_____
23. A card worth ten points	_____	_____

9 Practice • • • • • • • • • • • • • It's All in the Deck of Cards

Directions: Shuffle a deck of playing cards and answer the following questions.

1. If you were to pick a card from the deck, what are the possible outcomes?

   _____

   _____

   _____

2. Predict what would happen if you were to pick 10 cards from the deck.

   _____

   _____

   _____

3. Select ten cards from the deck. Color a box to show the card's category.

Both Odd										
Both Even										
One Odd/One Even										
Both Face Cards										
Both Cards Worth Ten Points										
Both Cards Worth Less Than Ten Points										
Consecutively Numbered										

4. Compare the results to your prediction. _____

   _____

   _____

 **More Than One Suit**

Directions: Circle the outcome that would be more likely and tell why.

1. Two cards of the same suit | Two cards of different suits

2. Two face cards | Two numbered cards

3. Two odd-numbered cards | An odd-numbered card and an even-numbered card

4. Two consecutively numbered cards | Two non-consecutively numbered cards

5. Two Aces | An Ace and a card with a value of ten

6. Two cards with different values | Two cards with one card having twice the value of the other card

7. Two cards with a product of one hundred | Two cards with a product of ninety

8. Three cards of the same suit | Three cards of different suits

9. Three cards in consecutive order | Three cards in random order

10. Three face cards | Three number cards

11. Three odd-numbered cards | Two even-numbered cards and one odd-numbered card

12. Picking four cards of the same suit | Picking two of four cards of the same suit

#3739 How to ... Data and Probability: Grade 3   40   © Teacher Created Materials, Inc.

10 Brain Teasers • • • • • • • • • • • • • • • • • • • Too Many Choices

Directions: How many possible combinations are there? Use multiplication to figure out the different possible combinations.

Example: Jamie is buying a car. She can buy a two-door or a four-door car. The color choices are red, white, and blue. How many different choices does Jamie have?

Body Styles	Colors	
Two-door	Red	Multiply the number of body styles by
Four-door	White	the number of color choices to find
	Blue	the answer.

2 (body styles) x 3 (color choices) = 6 choices   Jamie has six choices.

---

1. It's after-school sports time and the students have to pick out their shoes and socks. The students can wear high-top, low-top, or slip-on tennis shoes and knee socks, ankle socks, or sport socks. How many choices do the students have?

   Tennis Shoes   **Socks**

   _____ (shoes) x _____ (socks) = _____   The students have _____ choices.

2. At lunch, the students get to make their own pizza. The students pick out the pizza size (mini or small), the crust (thick or thin), and one topping (ham, pepperoni, or pineapple). How many choices do the students have?

   Pizza Size   **Crust**   **Topping**

   ____ (sizes) x ____ (crusts) x ____ (toppings) = ____   The students have ____ choices.

3. During cooking class, the students made cookies (peanut butter, sugar, or chocolate chip), cakes (vanilla, strawberry, or chocolate), and frosting (butter cream, banana, or chocolate). How many choices do the students have?

   Cookies   **Cakes**   **Frosting**

   ____ (cookies) x ____ (cakes) x ____ (frostings) = ____   The students have ____ choices.

© Teacher Created Materials, Inc.

10 Brain Teasers • • • • • • • • • • • • • • • • • Even More Choices

Directions: How many possible combinations are there? Use multiplication to figure out the different possible combinations.

1. The Smith family is building a house. Their choices are one- or two-story houses, the number of garages (two-, three-, or four-car), and a choice of one upgrade (swimming pool, sauna, or tennis court). How many possible combinations are there?

   Number of Stories   **Number of Garages**   **Extra Upgrades**

   ____ (stories) x ____ (garages) x ____ (extra upgrades) = ____
   The Smith family has _____ possible combinations from which to choose.

2. Brenda is making a sundae. She has to pick the ice cream flavor (chocolate, vanilla, strawberry, or rocky road), the sauce (hot fudge, chocolate, or caramel), the topping (walnuts, whipped cream, or cherry), and the cone (waffle or sugar). How many possible combinations are there?

   Ice Cream Flavors   **Sauces**   **Toppings**   **Cones**

   ____ (flavors) x ____ (sauces) x ____ (toppings) x ____ (cones) = ____
   Brenda has ____ possible combinations from which to choose.

3. Dilbert is looking at bikes. He needs to pick the bike's frame (mountain, racing, street), the kind of wheels (touring, racing, street), the number of speeds (three, five, ten, or twelve), and an accessory (water bottle holder, basket, tire pump, or kickstand). How many possible combinations are there?

   Frames   **Wheels**   **Speeds**   **Accessories**

   ____ (frames) x ____ (wheels) x ____ (speeds) x ____ (accessories) = ____
   Dilbert has ____ possible combinations from which to choose.

4. The local photography school holds a wide variety of classes. Students can pick from working with different cameras (Polaroid, digital, 35 mm, or pin-hole), with different kinds of film (black and white, sepia, or color), with different sizes of prints (3" x 5", 4" x 6", 5" x 7", or 8" x 10"), and with different themes (garden, action, babies, or animals). How many possible combinations are there?

   Cameras   **Film**   **Print Sizes**   **Themes**

   ____ (cameras) x ____ (film) x ____ (print sizes) x ____ (themes) = ____
   The students have possible ____ combinations from which to choose.

11 Technology • • • • • • • • • • • • • • • • • • Tables and Graphs

Direction: Learn more about tables and graphs by visiting the *Tables and Graphs* Web site at the address below. Write three facts about each kind of table or graph.

http://pittsford.monroe.edu/jefferson/calfieri/graphs/TabGraphMain.html

A. Tables

   1. _____

   2. _____

   3. _____

B. Bar Graphs

   1. _____

   2. _____

   3. _____

C. Column Graphs

   1. _____

   2. _____

   3. _____

D. Line Graphs

   1. _____

   2. _____

   3. _____

E. Circle/Pie Graphs

   1. _____

   2. _____

   3. _____

When would you use each kind of graph?

F. Tables: _____

G. Bar Graphs: _____

H. Column Graphs: _____

I. Line Graphs: _____

J. Circle/Pie Graphs: _____

11 Challenge • It's All in the Dots

Directions: Figure out the possible outcomes (adding the top number to the bottom number) for selecting a domino from a set of double-six dominoes.

1. The possible outcomes are the following: _____

2. What sum do you think will occur the most often? Why? _____
_____

Directions: At random, pick 12 dominoes. Make a bar graph below and record the sums on the chart.

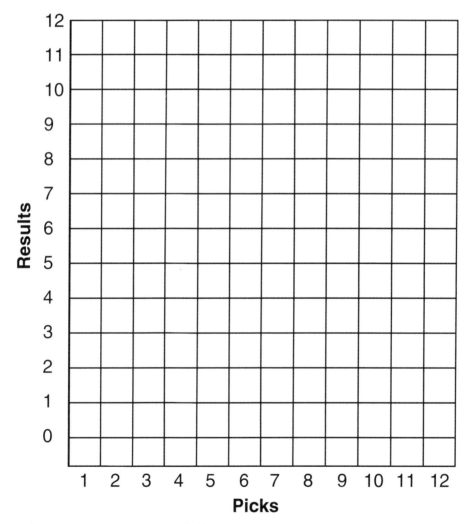

3. Compare the results to your prediction. _____
_____

4. If this experiment were repeated, would the results be the same? Why or why not? _____
_____

5. Find the mean, median, and mode.   Mean: _____   Median: _____   Mode: _____

12 Challenge •••••••••••••••••• Double-Line Graph

Directions: Use a nine-sided die to create a double line graph. Use a red pen to predict what number on which you think the die will land and a green pen to show the actual number on which the die landed. Record twenty predictions and rolls. How did you do?

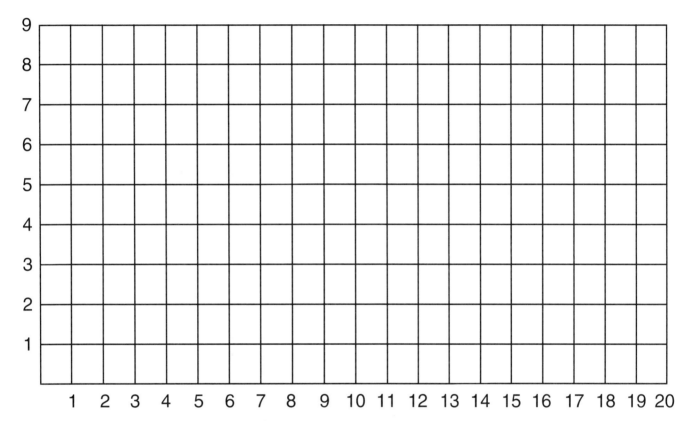

Directions: Analyze the results.

1. Out of twenty predictions, I was correct _____ times or _____ % of the time.

2. The mean (average) of the numbers rolled was _____ .

3. The median (middle number) was _____ .

4. The mode (number that was rolled the most often) was _____ .

5. The range (difference between the highest number and the lowest number) was _____ .

6. _____ odd numbers were rolled

7. _____ even numbers were rolled

8. _____ or _____ % of the numbers were odd.

9. _____ or _____ % of the numbers were even.

10. If this experiment were repeated, would the results be similar? Why or why not?

12 Challenge — Go on a Scavenger Hunt

Directions: Use the local newspaper to look for any articles or sections containing statistical or probability information.

Section	Page	Article Heading	Statistical or Probability Information About	Kind of Graph or Table Used

1. Which graphs were the easiest to read? Why? _____
2. Which graphs were the easiest to understand? Why? _____
3. Did any of the graphs explain how the data was interpreted? _____

#3739 How to ... Data and Probability: Grade 3

Answer Key

Page 6
1. Certain
2. Vary
3. Impossible
4. Vary
5. Impossible
6. Impossible
7. Vary
8. Vary
9. Impossible
10.–19 Answers will vary.
20. Impossible

Page 7
1. Unlikely
2. Likely
3. Unlikely
4. Unlikely
5. Unlikely
6. Likely
7. Unlikely
8. Likely
9. Likely
10. Likely
11. Unlikely
12. Likely
13. Unlikely
14. Unlikely

Page 8
1. U
2. U
3. E
4. U
5. L
6. U
7. U
8. U
9. U
10. L
11. L
12. E
13. E
14. L
15. L
16. L or U
17. L

Graph
Unlikely—7 or 8 boxes shaded
Equal Chance—3 boxes shaded
Likely—6 or 7 boxes shaded
Sample sentence: There were very few events that had an equal chance of happening.

Page 10
All of the boxes in the Red row are shaded. Sample answers shown below.
1. Red
2. It was the only jellybean in the bag.
3. Yes, because there is only one jellybean in the bag.
4. No, because there would be a chance of picking different colors.
5. No, because the other colors were left out of the sample population.
6. Using all of the jellybeans would make the experiment more accurate.

Page 11
Graphs and answers will vary.

Page 12
1. Answers will vary.
   Graph: Results will vary.
2. Answers will vary.
3. Answers will vary.
4. Answers will vary.
5. Sample answer: Yes, because the same set of jellybeans would be used.
6. Sample answer: Selecting jellybeans from a set of 45 jellybeans is most accurate because the other colors will have an equal chance of being picked.

Page 14
Sample Answers
1. Names of animals, numbers in increments of 5
2. Names of animals
3. Numbers in increments of 5
4. Each shaded box represents 5 animals.

5. 30 – 0 = 30
Circle the following questions: 7, 8, 11, 12

Page 15

	Tally Marks	Total Number
Baseball	|||||		6
Skiing	|||||	5
Swimming	||||	4

Bar Graph:
Baseball—6 boxes shaded
Skiing—5 boxes shaded
Swimming—4 boxes shaded
3. Sample statement: More classmates picked baseball as their favorite hobby.
4. Sample question: Which sport received the fewest number of votes?

Page 16
Step 2: 50, 50, 50, 60, 80, 80, 90, 90, 90, 100
Step 4: Sample Questions—What was the average number of stars made?, Which students made the same number of stars?, Who made more stars than Ursula?

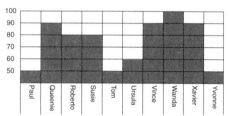

Page 18 Answers will vary.
Page 19 Answers will vary.
Page 20 Answers will vary.

Page 22

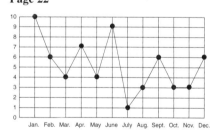

Page 23
Pie Chart #1

1. $2.00 ÷ $8.00 = .25 x 100 = 25%
2. $1.00 ÷ $8.00 = .125 x 100 = 12.5%
3. $4.00 ÷ $8.00 = .50 x 100 = 50%
4. Sample answer: Color each section of the pie chart a different color.

Pie Chart #2

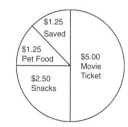

5. 50% or $5.00 for a movie ticket
6. 12.5% or $1.25 for pet food
7. 25% or $2.50 for snacks
8. 12.5% or $1.25 saved

Page 24

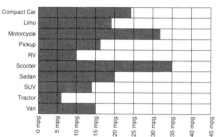

1. Sample sentence: The smaller vehicles seem to have better gas mileage.
2. Sample question: Which vehicle had the worst mileage?

Page 26
1. 581
2. 10
3. 105 – 14 = 91
4. 581 ÷ 10 = 58.1
5. Abby, Deandre, Fred, Gabby, Inez
6. Ben, Cassie, Erica, Hal, James
7. A bar graph can best represent this information. It will let the reader easily see the number of newspapers each newspaper carrier delivers.

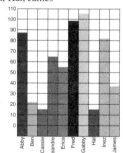

8. Names and numbers of papers
9. Sample sentence: Gabby delivers almost 50 more newspapers than Erica.
10. Sample questions: Who delivers fewer newspapers than Inez but more newspapers than Erica?, Which two carriers combined deliver fewer newspapers than James?

Page 27
1. 327
2. 5,510
3. 327; 427; 692; 725; 776; 921; 956; 1,033; 4,110; 5,510
4. 5,510 – 327 = 5,183
5. 776 + 921 = 1,697 ÷ 2 = 848.5
6. A line graph can best represent this information because it can track the number of pennies saved throughout the years.

© Teacher Created Materials, Inc.

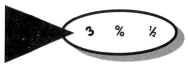

Answer Key

7. Number of pennies and the years
8. Sample sentence: There were almost five times more pennies saved in 1996 than in the year 2000.
9. Sample questions: What was the average number of pennies saved each year?, In which years were more than 1,000 pennies saved?

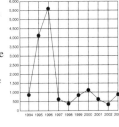

Page 28
1. 4
2. 10 – 4 = 6
3. A pie chart can best represent this information because a specific number of donuts are packed in each box.

4. Kinds of donuts and number of donuts
5. Sample sentence: There are more than twice as many chocolate donuts than old fashioned and plain donuts.
6. Sample questions: Which two donuts represent 1/3 of the donuts in the box? Which kind of donut represents 25% of the donuts in the box?

Page 30
1. 1, 2, 3, 4, 5, 6
2. 1/6
3. 1 ÷ 6 = .166 x 100 = 16.6%
4. Independent
5.–10. Answers will vary. Graphs will vary.

Page 31
Possible outcomes:

1-1	2-1	3-1	4-1	5-1	6-1
1-2	2-2	3-2	4-2	5-2	6-2
1-3	2-3	3-3	4-3	5-3	6-3
1-4	2-4	3-4	4-4	5-4	6-4
1-5	2-5	3-5	4-5	5-5	6-5
1-6	2-6	3-6	4-6	5-6	6-6

2. 1/6 x 1/6 = 1/36
3. 1 ÷ 36 = .02 x 100 = 2%
4. Independent
5.–7. Answers will vary.
8. No, because each roll of the dice is independent of the other rolls. Each combination of numbers has an equal chance of being rolled.

Page 32
Graphs will vary.

Page 34 Answers and graphs will vary.

Page 35
1. 1/2, 50%
2. 1/3, 33%
3. 1/4, 25%
4. 1/4, 25%
5. 0, 0%
6. 1/4, 25%
7. 2/8 or 1/4, 25%
8. 0, 0%
9. 2/5, 40%
10. 1
11. 2, 3, 4, 5, 6, 7, 8, 9
12. 1, 2, 3, 6, 7 There is an equal chance of landing on red.
13. 4, 5, 8 A smaller section or no section is shaded red when compared to the other color(s).

Page 36
1. Unlikely
2. Likely
3. Unlikely
4. Unlikely
5. Unlikely
6. Likely
7. Likely
8. Likely
9. Likely
10. Unlikely
11. Unlikely
12. Likely

Page 38
1. Ace, 1, 2, 3, 4, 5, 6, 7, 8, 9, 10, Jack, Queen, King
2. Unlikely
3. Likely
4. Likely
5. Unlikely
6. Likely
7. Likely
8. Unlikely
9. Picking a card with a value of ten
10. Picking a card with a value less than eight
11. Picking an odd-numbered card
12. Picking an even number
13. Picking an even-numbered card
14. Picking a card with a greater value than 5
15. 3/13, 23.08%
16. 10/13, 76.92%
17. 5/13, 38.46%
18. 8/13, 61.54%
19. 8/13, 61.54%
20. 4/13, 30.77%
21. 6/13, 46.15%
22. 1/13, 7.69%
23. 4/13, 30.77%

Page 39
1. Ace, 2, 3, 4, 5, 6, 7, 8, 9, 10, Jack, Queen, King for each suit or 52 possible outcomes
2. Answers will vary.
3. Graphs will vary.
4. Answers will vary.

Page 40
All reasons: Because there are more cards that fit the category.
1. Two cards of different suits
2. Two numbered cards
3. An odd-numbered card and an even-numbered card
4. Two non-consecutively numbered cards
5. An Ace and a card with a value of ten
6. Two cards with different values
7. Two cards with a product of one hundred
8. Three cards of different suits
9. Three cards in random order
10. Three number cards
11. Two even-numbered cards and one odd-numbered card
12. Picking two of four cards of the same suit

Page 41
1. Tennis shoes: High top, low top, slip-on
   Socks: knee, ankle, sport
   3 (shoes) x 3 (socks) = 9
   The students have 9 choices.
2. Pizza size: mini, small
   Crust: thick, thin
   Topping: ham, pepperoni, pineapple
   2 (sizes) x 2 (crusts) x 3 (toppings) = 12
   The students have 12 choices.
3. Cookies: peanut butter, sugar, chocolate chip
   Cakes: vanilla, strawberry, chocolate
   Frosting: butter cream, banana, chocolate
   3 (cookies) x 3 (cakes) x 3 (frostings) = 27
   The students have 27 choices.

Page 42
1. Number of stories: one, two
   Number of garages: two, three, four
   Extra upgrades: swimming pool, sauna, tennis court
   2 (stories) x 3 (garages) x 3 (extra upgrades) = 18
   The Smith family has 18 possible combinations from which to choose.
2. Ice cream flavors: chocolate, vanilla, strawberry, rocky road
   Sauces: hot fudge, chocolate, caramel
   Toppings: walnuts, whipped cream, cherries
   Cones: waffle, sugar
   4 (flavors) x 3 (sauces) x 3 (toppings) x 2 (cones) = 72
   Brenda has 72 possible combinations from which to choose.
3. Frames: mountain, racing, street
   Wheels: touring, racing, street
   Speeds: 3, 5, 10, 12
   Accessories: water bottle holder, basket, tire pump, kickstand
   3 (frames) x 3 (wheels) x 4 (speeds) x 4 (accessories) = 144
   Dilbert has 144 possible combinations from which to choose.
4. Cameras: Polaroid, digital, 35 mm, pin hole
   Film: black and white, sepia, color
   Print sizes: 3" x 5", 4" x 6", 5" x 7", 8" x 10"
   Themes: garden, action, babies, animals
   4 (cameras) x 3 (film) x 4 (print sizes) x 4 (themes) = 192
   The students have 192 possible combinations from which to choose.

Page 43 Answers will vary.

Page 44
1. 0-0, 0-1, 0-2, 0-3, 0-4, 0-5, 0-6, 1-1, 1-2, 1-3, 1-4, 1-5, 1-6, 2-2, 2-3, 2-4, 2-5, 2-6, 3-3, 3-4, 3-5, 3-6, 4-4, 4-5, 4-6, 5-5, 5-6, 6-6

2.–5. and graph: Answers will vary.

Page 45 Answers will vary.

Page 46 Answers will vary.